Burps Are Chemicals Too!

By Pamela J. Keck, Ph.D.
Illustrated by Thomas M. Plunk

Science Educators' Ink

Copyright © 2000 by Pamela J. Keck
All rights reserved. Printed in the USA.

Published by Science Educators' Ink,
P.O. Box 731, Glen Carbon, IL 62034

ISBN 0-9715268-0-X
LCCN 2001099254

Author and illustrator assume no responsibility for misuse of chemicals described in this book.

Acknowledgements
PJK expresses sincere thanks to: Anne Wallingford, Jenny Simonson Burke, Dennis, Dottie, Denise, Dawn, Courtney, LeAnne, Ron, Cindy, Brad, Laura, Hannah, Samuel, Mrs. Brasher's 1999 fifth grade class, Mrs. Harrison's 1999 second grade class, and as always, Don. TMP expresses sincere thanks to Deanne.

To Virginia, the original Dr. Dana (PJK)

For Mom (TMP)

As you go through this book, look for words chemists' use. The first time each is printed, it is underlined. How many have you seen before?

Lab

Molecule

Atom

Bond

Element

Formula

Polymer

Acids

Bases

pH Scale

Note: This book is meant to be custom colored by you. We recommend colored pencils.

Hi. I'm Betty the Beaker. I work in Dr. Dana's Science Laboratory. Or, as scientists like to say, <u>lab</u>.

Dr. Dana is a chemist. She studies chemicals in plants to find a cure for cancer. She is a scientist at a university and is called a chemistry professor. Chemistry is the study of what all things are made from and how they interact with each other and the environment.

While I like to help Dr. Dana in her chemistry lab by holding things...

What I really like, is going with Dr. Dana to schools for her Science Days.

Dr. Dana likes to talk to kids and teachers about science, especially chemistry. She places Florence the Flask, Thelma the Thermometer, and me at the head table. Chemists use us for mixing, measuring volume in milliliters (mL), and measuring temperature in degrees Celcius.

At the beginning of the science day activities, Dr. Dana asks some questions. The first usually is:

"What is matter?"

After a few seconds, someone might raise their hand and respond:
"Matter is anything that takes up space and has mass."

With a big grin, the chemistry professor happily responds **"YES!"** Dr. Dana reminds everyone that **matter can be a solid, a liquid, or a gas and mass is the amount of something.**

Then she asks:

"Which of the following things do you think are made of chemicals?"

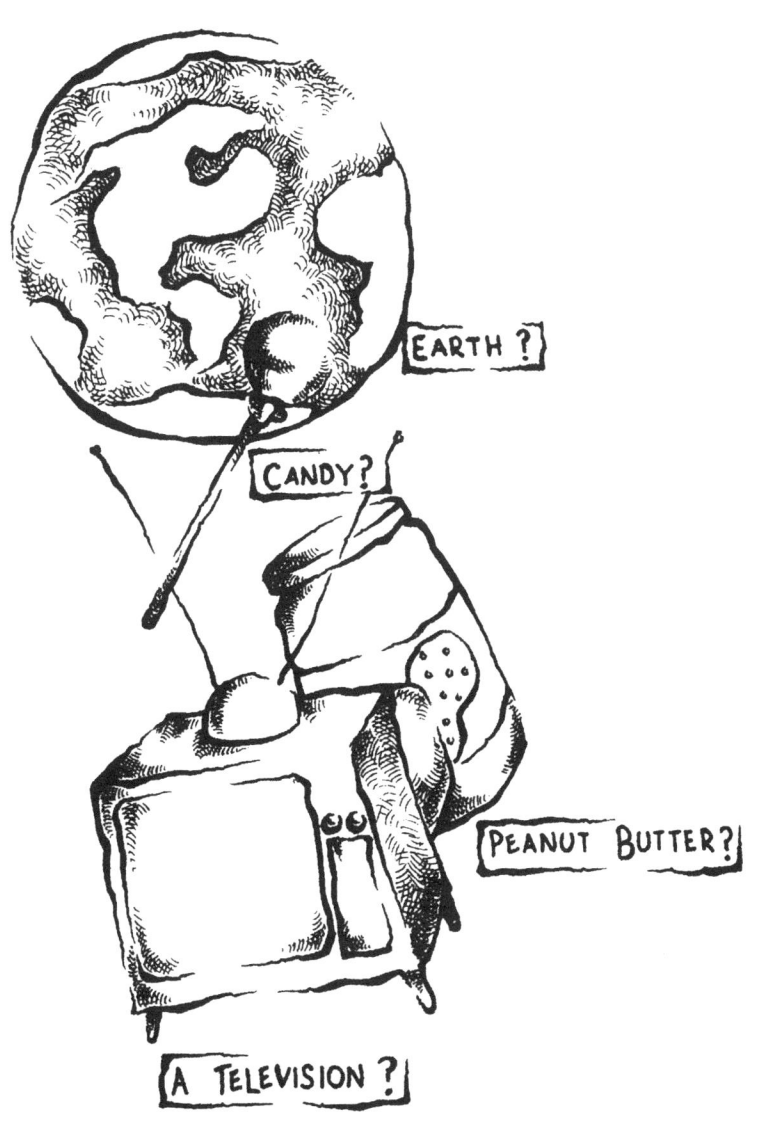

If you picked all the choices, you are correct! Because **all matter is made of chemicals** ....even BURPS!

Burps are made of chemicals that are gases. We can't see burps but can usually **HEAR** them! And with some people, **SMELL** them.

Burps come from our stomachs. While we eat, we swallow air. So, burps are made of air gases as well as gases from our food.

Air is made mostly of two chemicals called nitrogen and oxygen. Chemists might draw air to look like this:

```
_____

   N≡N            N≡N          molecules
                                  ←
        O=O           N≡N
_____
```

Each pair of N's and O's is called a <u>molecule.</u> They are joined together. From the picture, can you answer the following questions?

1. Which molecules do you think are the nitrogens?

2. Which molecules do you think are the oxygens?

3. Are there the same number of nitrogens and oxygens in the picture?

4. Are there the same number of lines between the N's as the O's?

Answers

1. The nitrogen molecules, are the ones with the N's.

2. The oxygen molecules, are the ones with the O's. Notice that each nitrogen molecule has two N's. Each "N" by itself is called a nitrogen <u>atom</u> and two nitrogen atoms make a nitrogen molecule. Each "O" is an oxygen atom and two oxygen atoms make an oxygen molecule.

    N is a nitrogen atom
    N≡N is a nitrogen molecule

    O is an oxygen atom
    O=O is an oxygen molecule

3. There are more nitrogen molecules in the picture because air has **more** nitrogen than oxygen. Air is *roughly* 3 parts nitrogen and 1 part oxygen. So, in air, there are 3 molecules of nitrogen for every 1 molecule of oxygen, just like in the picture.

4. There are three lines, called <u>bonds</u>, between the N atoms and only two bonds between the O atoms. Bonds hold atoms together to make molecules. Guess which molecule is harder to break apart into its individual atoms? (Nitrogen, the one with the more bonds).

Nitrogen is very important since it is found in every living thing. So, the nitrogen in the air is important for life. But, before living things can use it, the three bonds in nitrogen must be broken and new bonds formed. Nitrogen has to be bonded to oxygen or hydrogen before any living thing can use it.

Believe it or not, **lightning** breaks nitrogen-nitrogen bonds and then makes nitrogen-oxygen bonds. So, when it's thunderstorming and you see lightning, you can think about the fact that the nitrogen in the air is being changed (or 'fixed' as scientists say) so plants can use it.

Once a school principal asked Dr. Dana "How many different chemicals are there?" Dr. Dana was quiet for a long time. Florence, Thelma, and I weren't sure if she was counting, or if she didn't know the answer. Finally, she responded **"Since every thing is made out of chemicals, it is impossible to count the number of combinations of the different chemicals in the world."**

She continued, "There are 26 letters in the alphabet that make up all the words in the English language. Some letters are used much more often than others. When English words are made, the same letters can be used but in different amounts to make words with very different meanings. For instance, hot and hoot both use the letters H, O, and T. But the word hot uses one H, one O, and one T while the word hoot uses one H, **two** O's and one T. Hot and Hoot, as you know, have very different meanings!

There are about 100 different chemical <u>elements</u> that make up all the chemicals in the world. And just like letters in the alphabet, some of the elements are used much more often.

Scientists use one or two letters to abbreviate names of the elements. Some of the common elements are:

      H, hydrogen   C, carbon   Na, sodium
      S, sulfur   Ca, calcium   Fe, iron

Many more can be found in a Periodic Table (the Internet has lots of examples).

In the same way words are formed, different numbers of chemical element pieces, atoms, can be used to make chemicals with very different properties. For instance, two hydrogen atoms and one oxygen atom make H_2O, commonly called water. [Humans are about 2/3 water and ¾ of the Earth is covered with water; it's very important to us!] But, if we put two hydrogen atoms and two oxygen atoms together this is something entirely different, H_2O_2, called hydrogen peroxide, which is dangerous to drink.

The chemical words H_2O and H_2O_2 are called <u>formulas</u>. **Formulas tell us the number and type of each atom contained in different chemicals.**

Just like the same letters can be combined to form different words, the same words can be connected differently to make sentences with very different meanings. For instance, would it make any difference if:

Frogs ate worms.

OR

Worms ate frogs.

I think it would! Similarly, the same chemicals can be connected differently to make new substances with very different properties. The chemical molecule glucose, has the formula $C_6H_{12}O_6$ (this means each molecule has 6 carbon atoms, 12 hydrogen atoms and 6 oxygen atoms). A glucose molecule has six sides, like a honeycomb or hexagon. The lines are bonds between the carbon atoms (found at the points on the hexagon).

part of a glucose molecule

When many glucose molecules are hooked together side-by-side with oxygen atoms, it makes nutritious starch (found in potatoes) but hooked together at an angle makes cellulose or fiber (found in plants). **Our stomachs can break down (digest) starch. But it cannot digest cellulose; it passes through our bodies untouched.**

starch

cellulose

Starch and cellulose are both made of many molecules of glucose joined together differently.

When *many* molecules of the same substance are hooked together over and over, this is called a <u>polymer</u>. Poly means *many* and -mer means *unit*. This means that both starch and cellulose are polymers because they are made of **repeating units** (molecules) of glucose.

Dr. Dana likes to discuss slime, a polymer of polyvinyl alcohol. It is simply many vinyl alcohol molecules hooked together.

The students get to mix together polyvinyl alcohol with borax, a detergent found in the grocery store. When the two are shaken together, the mixture turns into a rubbery substance that can be bounced, squished, or poured onto a table and watched (recipe is found in the back of the book).

Polymers are used in many things. Plastics are made out of polymers. Different plastics have different properties. Toothbrushes, soda bottles, lunch boxes, milk jugs, garbage bags, racquetballs, and football helmets (to name just a few) are all made out of polymers.

A polymer you may have heard of before, and one that occurs in every living thing, is DNA. (short for deoxyribonucleic acid). The repeating units of this polymer are called A (short for adenine), T (short for thymine), G (short for guanine), and C (short for cytosine). Just like maps help us figure out where to go, the order these four units are hooked together is the 'map' that makes something a human, a plant, or an animal. Yep, that's right, you and bullfrogs each have DNA, but it's arranged a bit differently.

After discussing that all things are made out of chemicals, and the only thing that separates us from a bullfrog is the pattern of our DNA, Dr. Dana brings up the subject of <u>acids</u>, <u>bases</u>, and the <u>pH scale</u>. The pH scale is unlike any scale you've ever heard. It's not like a music scale, or the scale on a dinosaur or even a bathroom scale.

The pH scale contains numbers and tells chemists if they have an acid, a base, or a neutral chemical. If something is an acid it has a pH value of 0 to just below 7. The closer the number is to 0, the stronger the acid. If something is a base, it has a pH value of just above 7 to 14. The closer the number is to 14, the stronger the base. If something is neutral, it has a pH value of 7. pH means the 'power of hydrogen'.

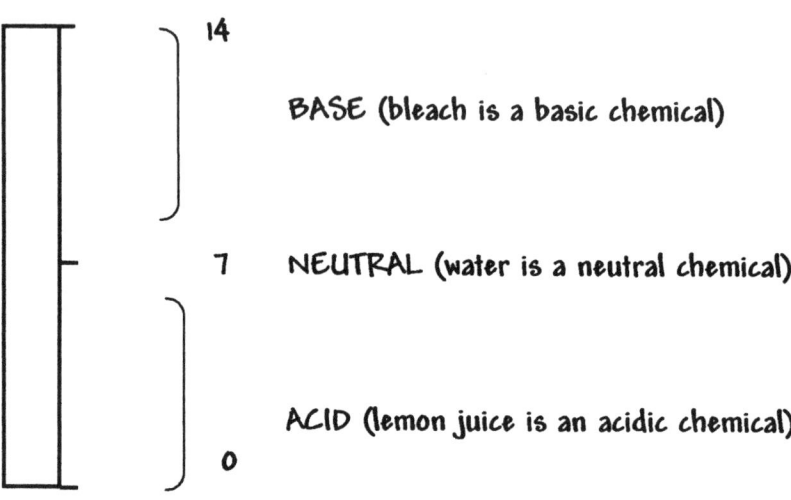

Dr. Dana uses red cabbage juice (from the grocery store) and Florence the Flask for this part of the show. The preparation of this juice is described in the back of this book. She explains that the liquid can be used to determine if a chemical is an acid, a base, or neutral. When the red cabbage juice is prepared in water (usually close to neutral pH), the color is blue.

She holds up Florence to demonstrate the starting color of the juice (blue).

She then explains that **acids** turn the blue juice PINK, and bases turn it either GREEN (a weak base) or YELLOW (a strong base). Dr. Dana then pours a bit of the juice into each of three beakers.

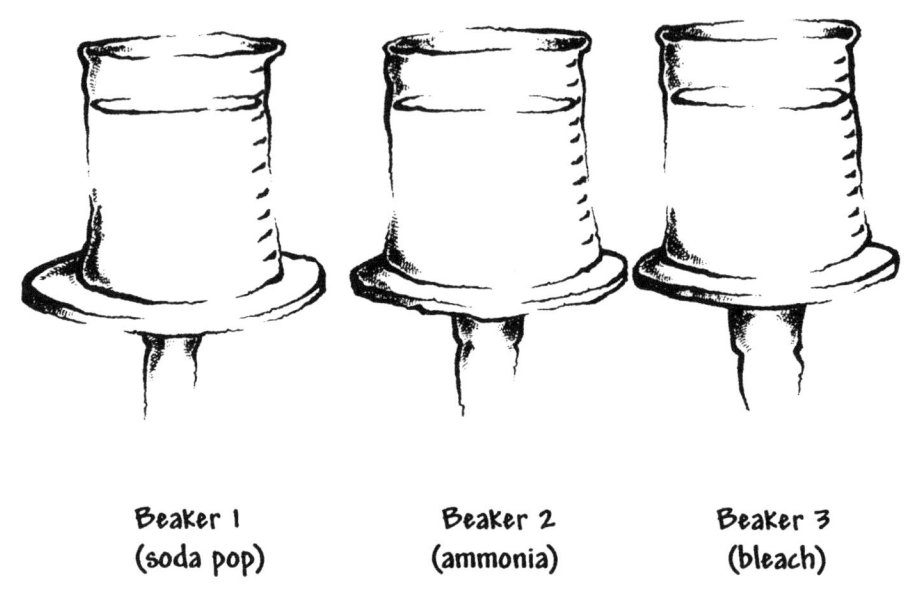

Beaker 1　　　　Beaker 2　　　　Beaker 3
(soda pop)　　　(ammonia)　　　(bleach)

Into beaker 1, she adds colorless soda pop to the cabbage juice and the solution turns PINK! Into beaker 2 she pours ammonia, and it turns GREEN! Into beaker 3, she pours bleach and it turns YELLOW! This means that soda pop is an ACID, ammonia is a weak BASE and bleach is a strong BASE. In fact, most household cleaners are bases and most bases are slippery (as are ammonia and bleach).

Students test chemicals to see if they are acids, bases, or neutral. Dr. Dana brings color-changing modeling 'dough' made with red cabbage juice (recipe is found in the back of the book). The students squish the dough with cream of tartar (tartaric acid), smashed up aspirin (acetylsalicyclic acid), baking soda (sodium bicarbonate), or powdered sugar (sucrose). They all get to take home their chemical dough samples in polymer bags (plastic bags).

After the students learn that everything is made out of chemicals, that slime is a polymer and that most household cleaners are bases, the Science Day is complete. We are ready to go back to the lab and search for chemicals to treat cancer!

Here are some questions that you should be able to answer after reading this book:

1. What is matter?
2. Is there any type of matter that is not a chemical?
3. What are the three common states of matter?
4. What chemical is the major component of air?
5. What holds atoms together?
6. What does 'fixing nitrogen' mean?
7. How many different chemicals are there?
8. What is a chemical formula?
9. What chemical are starch and cellulose made from?
10. What is a polymer?
11. What polymer, discussed in this book, is present in you and a bullfrog (all living things)?
12. What does pH stand for?
13. An acid has a pH value below ____ and a base has a pH value above ____?

Recipes
(young children will need assistance)

Cabbage Juice

1. Buy red cabbage at the grocery store. Usually it is found right by the green cabbage.
2. Cut ½ of the cabbage into pieces about the size of a dime.
3. Pour into a pot and cover with H_2O, water, plus about another 240 mL (a cup) of water.
4. Boil for about 8 minutes.
5. Allow to cool. Let your cat smell this for some fun.
6. Drain off the liquid (should be blue) and store in the refrigerator for several days or freeze and store for longer periods (several months).

Color Changing Dough

240 mL (1 cup) of sodium chloride (table salt)

240 mL of corn starch (glucose polymer)

240 mL of cabbage juice

12 mL (1 tablespoon) of oil (triacylglycerides)

Stir the two solids together then add the two liquids. Heat over medium heat until the mixture starts to stick together; heat an additional 3 minutes, while stirring. Remove from heat, let cool, and roll into balls approximately 2.5 cm in diameter (about the size of a quarter).

Slime

First prepare:

4% solution of Borax (40-mule team Borax, available in the detergent section of the grocery store)

4% solution of polyvinyl alcohol (available from Flinn Scientific)

4% solutions are made by mixing 4 grams of powder and diluting to a final volume of 100 mL. This can be scaled up proportionately (for example, 40 grams diluted to 1000 mL, if desired). The mixture must be heated in a microwave (this can be dangerous since it gets very hot) a little at a time and stirred until it dissolves.

Once the solutions have been made and cooled, the polyvinyl alcohol can be colored by adding food coloring. To make slime, mix about 4 mL of the 4% borax with about 30 mL of the 4% polymer; shake in a closed tube for about 5 minutes.

Order Form

Send to (please print):

Name _____

Address _____

City _____ State _____ Zip _____

Telephone number: _____ email _____

Quantity	Item and Unit Cost	Cost
	Burps are Chemicals Too! ($6.95 ea.)	
	Shipping Costs (see below)	
	Total	

Shipping/handling costs:
1-5 copies $2.50
5-10 copies $4.50
10-more copies $6.50 (large orders will be billed appropriately)

Return to:
Science Educators' Ink, P.O. Box 731, Glen Carbon, IL 62034

Questions? Call: 618-659-9714